The successful treatment of a patient with psychosis who carries SNPs that are significant markers of schizophrenia in Irish high density schizophrenia families and who has MTHFR deficiency.

Author: John Neville

Contact details: jwilljohnwill@gmail.com

Keywords/phrases: Schizophrenia;   Irish   high   density schizophrenia families; SNAP25; MTHFR; IL3.

Published: 23.06.2015

Addendum added: 12.09.15

Publisher: Medresind

ISBN: 978-1-326-32255-7

First edition

Copyright: John Neville 23.06.2015

ISBN: 978-1-326-32255-7

Abstract

A proband who has a familial disorder which causes psychosis and who carries SNPs associated with schizophrenia in Irish high density schizophrenia families was successfully treated for his psychosis, anxiety and panic with supplements and dietary changes. These treatments may be replicable in other patients who carry some of the same SNPs as the proband. The proband's responsiveness is analysed in light of a hypothesis that proposed that the proband's familial disorder could result in abnormal TP53 expression. Evidence is discussed which suggests that the proband's C677T (TT) MTHFR deficiency may have impacted on his TP53 metabolism, providing a partial explanation for his supplement responsiveness. The proband's dietary responsiveness is analysed in light of his carrying SNAP25 SNPs that are also found in Irish high density schizophrenia families. It is suggested that the proband's dietary responsiveness may be because SNAP25 plays a key role in axonal growth, stimulates neurite sprouting and/or is involved in the exocytosis of biogenic amines. The proband's calcium phosphate responsiveness is also discussed, as is a possible relationship with Anderson's disease. It is proposed how the proband's disorder could create vulnerability to parasitic and fungal infections. Finally, it is speculated that IL3 might play a role in the control ERE function. In which circumstances, in the case of the proband, a unifying hypothesis could be that IL3 SNPs may interrupt the presumed estradiol up-regulation of both P5CS and arginase.

List of abbreviations:

ACSL6 - Acyl-CoA Synthetase Long-Chain Family Member 6.

AChE – actetylcholinesterase.

cGMP – Cyclic Guanosine Monophosphate.

CSF2RB - Colony-Stimulating Factor-2 Receptor, Beta, Low-Affinity.

ERE - Estrogen Response Elements

GAD1 - Glutamic Acid Decarboxylase 1.

GS – L-Glutamate-5-semialdhehyde.

GRN – Progranulin.

IL3 - Interleukin 3.

IHDSF - Irish high density schizophrenia families.

iNOS – Nitric Oxide Synthase 1. -

MTHFR - Methylenetetrahydrofolate reductase.

NO – Nitric Oxide.

OCTN1 - Solute Carrier Family 22 (Organic Cation/Zwitterion Transporter) Member 4.

OD – Ornithine Decarboxylase.

PKG - cGMP dependent protein kinase.

PMP22 - Peripheral Myelin Protein 22.

PRODH - Proline Dehydrogenase.

P5CA – Pyrroline-5-carboxylic Acid.

P5CR - Pyrroline-5-carboxylate Reductase.

P5CS - 1-Pyrroline-5-carboxylate Synthetase.

RAI1 – Retinoic Acid Induced 1.

SMS - Smith Magenis Syndrome.

SNAP25 - Synaptosomal-Associated Protein of 25 kDa.

SNP - Single Nucleotide Polymorphism.

TP53 - Tumour Protein 53.

Introduction

The proband has suffered from psychosis along with symptoms of P5CS/P5CR disorders and symptoms of 17p11.2 disorders. He reports that he has a family history on the paternal side of psychosis that is comorbid with symptoms of P5CS/P5CR disorders and symptoms of 17p11.2 disorders.

A previous paper proposed a hypothesis (the 'hypothesis') that suggested that SNPs located at IL3/ACSL6 at 5q31.1 which are significantly associated with schizophrenia in IHDSF and which are located in or around EREs result in or from the failure or partial failure of one or more of these EREs to activate P5CS/a pathway from glutamate to P5CS to P5CR to proline (the 'proline pathway'). This would then have flow on effects on other pathways and could result in compensatory TP53 activity to up-regulate the proline pathway.[1]

The above hypotheses might appear to run contrary to the prevailing view that schizophrenia is a highly heritable disorder where 'genetic risk is conferred by a large number of alleles, including common alleles of small effect.'[2] However, it is explained here why this apparent inconsistency may not be what it seems.

The proband's responsiveness to supplements and dietary changes are also examined and a potential explanation for his responsiveness is provided.

Treatment history

At about age seven the proband underwent inconclusive hospital investigations for cyclic diarrhoea, fever and lethargy which he had suffered from since birth. These symptoms were never explained and whilst they improved significantly around age nine, they did not completely resolve.

After the initial onset of psychotic symptoms in his thirties the proband's psychotic episodes gradually became increasingly frequent, more severe and of longer duration. The episodes were jealous and later persecutory and were accompanied by severe anxiety, panic and insomnia. He additionally felt 'brain fog', apathy, disinhibition, loss of empathy, inertia, impaired decision making, metal rigidity and inflexibility, and emotional blunting.

When he first sought psychiatric treatment he was commenced on olanzapine. This was effective in allowing him to sleep but in other respects it was largely ineffective. He reported that his anxiety and psychotic symptoms were dulled down when he took olanzapine, but that taking olanzapine was accompanied by 'brain fog', a sense of stupefaction, deteriorating memory problems, continuing intermittent pressure palsies, significantly deteriorating lower limb stiffness and peripheral neuropathy. The presence of the latter was subsequently confirmed by nerve conduction studies.

After a period of treatment, under the supervision of his doctor the proband ceased taking olanzapine. Within a few days he suffered a relapse into severe insomnia, panic, anxiety and paranoia.

Before recommencing olanzapine the proband took Blackmore's Executive Stress group supplement (the 'group supplement') for reasons unconnected to his psychiatric illness. He reports that significant recovery from his psychiatric illness took place within twenty-four hours of his taking the group supplement. He subsequently noted a repeating pattern of relapses within about a month of going off the group supplement and recovery within twenty-four hours of taking the group supplement. He reported that the group supplement was significantly more effective as a treatment for his psychosis than olanzapine.

Initially he took fish oil and probiotics along with the group supplement but he subsequently noted that the group supplement was effective in treating his psychosis in the absence of the fish oil and probiotics.

The proband's ongoing lose stools/diarrhoea had worsened in his late thirties and forties. However, whilst on the group supplement he also found he had significantly reduced lose stools/diarrhoea. His anxiety significantly improved on the group supplement whilst his night terrors and panic almost completely resolved. He had an ongoing sense of malaise and ongoing concerns about his memory. He continued to suffer from intermittent pressure palsies whilst his stiffness and peripheral neuropathy continued to deteriorate. Also, whilst his chronic severe insomnia resolved, he suffered once again from daytime sleepiness. He felt that he regained a real sense of perspective as to the events triggering his episodes. He continued to have mild occasional overvalued thoughts that were significantly less frequent and less severe than those that he had on olanzapine and which he found relatively easy to process. He continued to feel mild anxiety, 'brain fog', apathy, disinhibition, loss of empathy, inertia, impaired decision making, metal rigidity and inflexibility, and emotional blunting. Nevertheless the group supplement appeared to be such an effective treatment that the proband no longer met criteria for the diagnosis of a psychiatric illness.

The proband trialed each of the ingredients (excluding potassium monophosphate which he was unable to source) of the group supplement separately. He reported that his psychiatric symptoms responded well to lecithin, avena sativa, folic acid and calcium phosphate when taken individually and to a combination of B vitamins (B1, B2, B3, B5, B6 and B12) when these B vitamins were taken together. However, these individual supplements and the combination B vitamins were not as effective as the group supplement. It is noted here that each of the combination B vitamins was also trialed separately, including B12, and were not found to be effective.

The proband subsequently trialed the strict elimination diet utilised by Royal Prince Alfred Hospital in Sydney. His elimination diet additionally excluded legumes, mussels and all cooking oils except rice bran oil. The former were excluded because the proband and it is believed his father both reported severe reactions to lentils; the latter were excluded because the proband reported that cooking oils triggered paranoia in him. Whilst on the elimination diet until he completed blind and food challenges the proband did not take the group supplement.

Following blind and food challenges, the proband was diagnosed with multiple food intolerances, including intolerance to salicylates, lactose, amines, nitrates, antioxidants, colours, propionates and preservatives. He correctly identified the placebos in blind challenges for all the above. Diagnosis was confirmed by food challenges.

He reports that his night terrors, paranoia, anxiety, panic, insomnia, stiffness, peripheral neuropathy and pressure palsies are significantly improved on the elimination diet and that breaking the diet results in his suffering symptoms that include peripheral neuropathy, pressure palsies, night terrors, paranoia, anxiety, panic and/or insomnia. His diarrhoea is not improved by the elimination diet, although some foods such as mussels appear to cause severe diarrhoea.

When on the elimination diet he remains concerned about his memory and has an ongoing sense of mental malaise along with occasional significantly milder overvalued thoughts that he is able to process. Again, he continues to feel mild anxiety, 'brain fog', apathy, disinhibition, loss of empathy, inertia, impaired decision making, metal rigidity and inflexibility, and emotional blunting. Salicylates appear to trigger his peripheral neuropathy and pressure palsies a day or two after eating the salicylates.

Since the proband has been on the elimination diet he has become more sensitised to foods. He notes that in addition to responding very favourably to the group supplement he currently also has negative food intolerance responses to the group supplement and to avena sativa and lecithin when taken separately.

The proband's treatment regime

The proband usually takes Cenovis folic acid and Cenovis multivitamin and minerals, and he occasionally takes Blackmore's calcium phosphate. Additionally, he follows a strict elimination diet. He has made other lifestyle changes, such as seeking less stressful employment. His psychiatric symptoms are kept under control with this regime, but his diarrhoea is unaffected. He remains vulnerable to stressors and even to mild alterations to his diet/treatment regime.

Recently the proband has trialed taking potassium chloride with the above and early indications are that this may play a role in improving his diarrhoea. When he has additionally taken fatty acids with the potassium chloride, particularly in the form of flax seed oil, there appears to be further and significant improvement in his diarrhoea. However, taking flax seed oil is accompanied by other food intolerance symptoms.

The proband's abnormal test results and information on the disorder's symptoms

The proband's abnormal test results and detailed information on the symptoms and conditions found in the proband and his family members are found in a previous paper.[3]

Methylenetetrahydrofolate reductase deficiency

MTHFR is reported to catalyse the reduction of 5, 10-methylenetetrahydrofolate to 5-methyltetrahydrofolate, the main circulatory form of folate and carbon donor for the re-methylation of homocysteine to methionine.[4] The MTHFR gene is located at 1p36.3 and the MTHFR C677T (TT) polymorphism results in reduction of MTHFR activity by about 70%.[5] [6] Homocysteine may be high[7] and methionine low in MTHFR deficiency.[8]

Some case reports appear to suggest that MTHFR deficiency can of itself be a cause of schizophrenia.[9] [10] However, large scale studies have not provided consistent evidence of the involvement of the T-allele of the C677T polymorphism in schizophrenia, for example a meta-analysis of studies performed in 2007 was generally supportive[11] whilst a study dated 2012 was not.[12]

MTHFR deficiency is often treated by folic acid, B6, B12[13] [14] [15] and/or betaine.[16] When B12 is used as part of a treatment protocol for MTHFR deficiency where psychosis is present, unless B12 deficiency is ruled out before supplementation with

B12 commences, it is not always possible to rule out B12 deficiency as being the underlying cause of the psychosis.

Interestingly, Ding et al[17] examined the C677T polymorphism in connection with therapy related myelodysplasia or acute myeloid leukemia. They detected a synergistic effect between TP53 and MTHFR and proposed a model to explain the interaction between TP53 and MTHFR in which reduced MTHFR activity is associated with chromosomal aberrations during DNA repair. Other research suggested that the C677T polymorphism may provide a protective effect in reducing P53 oxidative damage in hepatocellular carcinoma probands.[18] Also, reduced MTHFR activity has been associated with a decreased risk of acquired mutations within the P53 gene occurring at CpG sites.[19] As such, it does appear that there is an important, albeit as yet incompletely understood, relationship between the C677T SNP and TP53.

The proband is homozygous for the MTHFR C677T (TT) polymorphism. However, the reported cases where treatment of MTHFR deficiency has been an effective treatment for a patient's psychosis have generally been cases of severe deficiency. The proband's test results do not indicate severe deficiency: his homocysteine, tested after his recovery from psychosis, was in the normal range (9.5 range 6-14 µmol/L); his methionine, also tested after his recovery from psychosis, was in the normal range (14, 16, 20, 21, 21, 22, 23 range 7-60 aemol/L); whilst his B12, which was tested when he was acutely ill, was also in the normal range (357 lower limit >180pmol/L) and would remain close to the normal range even if, as some experts suggest, the lower limit for B12 was doubled.

It is noted here that a study has been performed on the levels of B12, folic acid and homocysteine in children and adolescents with psychosis who also have the C677T polymorphism in which the levels of B12 and folic acid were found to be within normal limits whilst the average level of Hcy was 11.94 ± 5.6 µmol/l for patients with schizophrenia verses 6.80 ± 2.93 µmol/l in a control group.[20]

Notwithstanding the proband's homocysteine, methionine and B12 results, a significant disruption of the proband's homocysteine metabolism may be indicated because his serine was repeatedly low when he was on the group supplement or had been off it for a short period but normalised when he went off the group supplement for a lengthy period.

Dudman et al[21] reported a similar result in renal transplant patients treated with B12, B6 and folic acid and suggested that this could have resulted from the requirement for serine as a source of methyl carbon atoms in the methylation of homocysteine by N5-methyltetrahydrofolate and as a substrate in the cystathionine B-synthase reaction. It may be that the proband's low serine may have resulted from B12/B6/folic acid supplementation triggering the diversion of homocysteine to cystathionine.

Since MTHFR deficiency may be treated by folic acid, B6 and B12,[22] [23] [24] MTHFR deficiency could in part explain the proband's group supplement responsiveness and his separate folic acid and combination B vitamin responsiveness. However, he trialed betaine and found it ineffective, suggesting that in his case it may be the level of MTHFR activity which is significant and not the product of MTHFR activity.

It should be noted that TP53 is linked to colon cancer. Also that folic acid supplementation may be contraindicated in some cases where there is a risk of colon cancer.[25]   The proband has had colon polyps removed.

Phenylalanine hydroxylase, acetylcholine and MTHFR

Phenylalanine hydroxylase converts phenylalanine to tyrosine. The proband carries the phenylalanine hydroxylase rs1522305 (CC) SNP for which the common allele G is reported to have significant association with schizophrenia.[26] The proband carries at least 439 other SNPs on the phenylalanine hydroxylase gene, none of which are heterozygous.

Interestingly, given that the proband and his family suffer delusions and not hallucinations, when tested for their impact on five schizophrenia symptom factors (delusions, hallucinations, mania, depression, and negative symptoms) in a subset of the IHDSF, the phenylalanine hydroxylase 232 bp microsatellite allele was reported to demonstrate significant association with the delusions factor.[27]

It has been reported that:

1. Folic acid has been implicated in the conversion of phenylalanine to tyrosine and that tetrahydrofolic acid, the folic acid derivative required for the production of MTHFR, can substitute for the natural cofactor of phenylalanine hydroxylase in vitro;[28]
2. That in MTHFR C677T carriers AChE activity was significantly higher and decreased to normal levels after therapy with folic acid and that in an in vitro study incubation of homocysteine-activated membrane AChE from controls with phenylalanine resulted in restoration of AChE activity, but failed to reverse the stimulated

enzyme in hyperhomocysteinaemic MTHFR C677T subjects before therapy;[29] and

3. That when reconstituted in liposomes OCTN1 (SLC22A4) which is located at 5q31.1 (the location strongly implicated in the causation of the proband's disorder) catalyzes acetylcholine transport which is defective in the mutant L503F associated with Crohn's disease.[30]

SNAP25 and food intolerance

For the purpose of this paper, it is important to note that it has been demonstrated that SNAP25 plays a key role in axonal growth[31] and that SNAP25 stimulates neurite sprouting.[32] Axonal growth and dendritic/neurite sprouting are thought to be linked to the causation of food intolerances. Fanous et al[33] suggested that SNAP25 could be a susceptibility modifier gene in schizophrenia.

It has been reported that over-expression of dysbindin induced the expression of SNAP25 and increased extracellular basal glutamate levels and release of glutamate evoked by high potassium,[34] whilst suppression of dysbindin can result in an increase of the expression of SNAP25.[35] It has also been reported that dysbindin is required for the stabilisation of dendritic protrusions[36] and that SNAP25 plays a key role in the exocytosis of biogenic amines by neurons and endocrine cells.[37]

The proband carries a number of SNAP25 SNPs that Fanous et al found to be associated with schizophrenia in IHDSF and he also carries various dysbindin risk SNPs including rs760761 which has a joint effect with the 5q31.1 SNP rs31400 which he also carries.[38]

Calcium homeostasis, calcium channels and the relationships between calcium and TP53

There is a significant relationship between calcium and TP53. For example, it is reported that:

1. p53 and calcium signaling are inter-dependent and show both synergistic and antagonistic effects on each other in the cellular environment;[39]
2. The calcium-binding protein S100B down-regulates p53 and apoptosis in malignant melanoma;[40] and
3. High doses of calcium activate nitric oxide synthase, synthesizing nitric oxide which then downregulates Mdm2 and influences drastically the p53-Mdm2 network regulation.[41]

Corradini et al[42] summarised the significant amount of evidence behind the view that SNAP25 modulates voltage-gated calcium channels. Whilst Zhang et al[43] report that Ca2+-dependent synaptotagmin binding to SNAP-25 is essential for Ca2+-triggered exocytosis.

Homer 2 controls calcium homeostasis and the Homer 2 SNP rs869498 shows association with schizophrenia in an Irish schizophrenia sample.[44] The proband carries rs869498(CC).

It has also been reported that cGMP binding activates PKG, which phosphorylates serines and threonines on many cellular proteins, and that the proteins that are modified by PKG commonly regulate calcium homeostasis.[45] It is noted here that the hypothesis implicates abhorrent cGMP, PKG and iNOS related activity in the aetiology of the proband's disorder.[46]

Two of the SNP that have significant genome wide associated with schizophrenia carried by the proband[47] are located at

CACNA1I *(t*he pore forming alpha subunit of the Cav3.3 T-type calcium channel) which is located at 22q13.1.

Mattson's[48] suggestion of risks associated with calcium supplementation in Parkinson's disease is noted and is something which treating doctors should consider.

Anderson's disease (see addendum 1)

Anderson's disease which results from defects at SAR1B at 5q31.1 causes peripheral neuropathy, diarrhoea, lipid abnormalities and can cause hyporeflexia.[49]

The proband's known SAR1B SNPs are i5003997(CC), i5003999(TT), rs11749469(GG), rs2305049(GG), i3003109(CC), i5003998(CC) and i6023688(TT).

He has suffered from diarrhoea, hyporeflexia, peripheral neuropathy and different lipid abnormalities to those typically found in Anderson's disease. Diarrhoea is one of the symptoms that appears to be comorbid with psychosis in the proband's family. There is a suggestion that other family members with psychosis have also suffered peripheral neuropathy and it may be notable that hyporeflexia is also present in some family members.

It was previously proposed that since cyclic guanosine monophosphate (cGMP) is strongly associated with diarrhoea and since there appears to be a relationship between SAR1 and soluble guanylate cyclase which is directly connected to cGMP, the existence of a relationship between SAR 1 and cGMP might account for the diarrhoea in Anderson's disease and in the proband's familial disorder.[50]

Sec12 is a prolactin regulatory binding-element protein. The proband's prolactin test result was normal. It has been reported that mSec12 is essential for activation of SAR1.[51] It has also been reported that the crystal structure of Sec12 reveals that a single potassium ion stabilises a K loop and that bound potassium is essential for optimum guanine nucleotide exchange activity in vitro.[52]

Anderson's disease is in many respects effectively treated with a low-fat diet supplemented with lipid soluble vitamins A and E and essential fatty acids.[53] It is pointed out above that fatty acids, which the proband has taken with vitamins A and E in the Cenovis multivitimin, with potassium chloride and with his usual treatment regime, appear to significantly improve his diarrhoea, whilst at the same time resulting in food intolerance reactions. Also, early indications are that his diarrhoea may be partially responsiveness to potassium chloride.

Glutamic acid decarboxylase.

Hettema et al[54] identified six SNPS, rs2058725(A/G), rs3791851(A/G), rs2241165(A/G), rs769407(C/G), rs12185692(A/C) and rs3791850(C/T) in the *GAD1* region, that suggested association with anxiety disorders, major depression and neurotacism. The proband carries rs2058725(CT), rs3791851(CT), rs2241165(CT), rs769407(CG), rs12185692(CC) and rs3791850(AG).

There is a suggestion that the proband's high urinary glutamic acid/creatinine ratio results may correlate with when the proband was taking the group supplement. There is also a suggestion that his normal blood taurine results may have correlated with his taking the group supplement whilst his low blood taurine results may have correlated with when he was off supplements. However, insufficient records were kept to be sure there was such a correlation for either his blood taurine or his urinary glutamic acid.

Interleukin 3 and avena sativa

IL3 is significantly associated with inflammation. There is evidence suggesting that oat extracts exhibit anti-inflammatory activity.[55] Anti-inflammatory activity of avena sativa might explain the proband's responsiveness to avena sativa. However, after going on the elimination diet the proband noted food intolerance reactions to avena sativa as well as to lecithin. It would be interesting to know whether or not avena sativa plays a modulatory role in IL3 activity.

Jejunal villus heights and starvation

There is evidence that a proline to ornithine decarboxylase to polyamine pathway could be an opposing pathway to the proline pathway and that there could be cortisol induced over-expression of this polyamine pathway in the proband.[56] The cortisol stimulation of this polyamine pathway in piglets was reported to be associated with increased jejunal villus heights by 13% but had no significant effect on jejunal crypt depth or lamina propria depth.[57] Starvation results in decreased jejunal villus heights in rats[58] and is generally associated with jejunal mucosal atrophy.

A possible relationship between parasites and psychosis (see addendum 2)

As noted above the hypothesis suggested that the proband could have over-expression of ornithine decarboxylase, as well as iNOS.[59] IL3 which is heavily implicated in the proband's disorder is a type of cytokine.

There is some evidence that parasitic infestations, particularly of Blastocystitis Homonis and Microsporidia, are more prevalent in chronic psychiatric patients.[60] Some experts suggest that the parasite toxoplasma could be a prime candidate in the causation of schizophrenia.[61] African sleeping sickness is a parasitical infection where the parasite, trypanosomiasis brucei, requires polyamines for survival and can be treated with an ornithine decarboxylase inhibitor. It is reported that Blastocystisis ST7(B) may have evolved to downregulate NO production including via iNOS in order to evade host defences.[62] It has also been reported that Blastocystis could induce the expression of various proinflammatory cytokines.[63]

Symptoms of parasitic infections such as fatigue, GERD or symptoms of GERD, diarrhoea, poor coordination, numbness, vomiting, abdominal pain, sleep disturbance and inability to put on weight are found in some of the proband's family members who have suffered from psychosis. The proband's wife has been diagnosed with Blastocystitis Homonis. Stool sample results are awaited for the proband but it would not be surprising if he carried one or more parasitic infections.

It is noted here that increased plasma agmatine levels have been reported in schizophrenic patients[64] and that the hypothesis proposes that abnormal agmatine metabolism is implicated in the proband's disorder.[65] It has been reported that in the fungi Fusarium graminearum, which causes wheat blight, agmatine differentially regulates a large number of fungal genes.[66]

Arginase, estradiol, IL3 and TP53

It has been reported that homogeneous interleukin 3 enhances arginase activity in murine hematopoietic cells.[67]

There is evidence that arginase activity is modulated by estradiol[68] [69] and that there is complex crosstalk between P53 and arginase.[70] It has been reported that in in vitro whole blood cultures 17beta-estradiol results in significant decreases the spontaneous secretion of IL-6.[71] However, it has also been reported that IL-3 release was significantly depressed after trauma-hemorrhage in vehicle-treated mice, whereas these functions were maintained in 17beta-estradiol -treated mice.[72]

Luo et al.[73] found that IL3 activated both estrogen receptors, but that estrogen didn't directly regulate the expression of IL3. They also found that the C to T change at SNP rs31480 could change the binding affinity of transcription factor SP1 and influence the expression of IL3. It is recalled that estrogen receptor binds to estrogen response element and transactivate gene expression.[74]

Conclusions

It would be interesting to know if any patients with schizophrenia or psychosis alone linked to IHDSF have:

1. Symptoms of P5CS/P5CR disorders and 17p11.2 disorders; and
2. Low ornithine and/or other biomarkers that the proband has.

If so, do any of these symptoms/results correlate with treatment responsiveness to supplements and/or the elimination diet?

It is unlikely that in this case the C677T polymorphism was causal of the proband's psychosis, but it appears probable that in the proband the C677T polymorphism modifies the effect of the underlying disorder and that thus his psychosis was treated when his MTHFR deficiency was treated. His reporting a good response to treatments that apply to this polymorphism before he was aware he had the polymorphism appears significant.

If his abnormal MTHFR metabolism connected closely with his underlying familial defect, as could be the case given the likely relationship between C677T polymorphisms and TP53 detailed above, this could explain his responsiveness. Perhaps in the proband his C677T polymorphism through its relationship to TP53 plays a significant role in aggravating the consequences of the familial disorder.

Vitamin deficiency can cause psychosis and there are some rare disorders that cause psychosis that are vitamin responsive, for example Hartnup disorder. However, there are a large numbers of anecdotal reports suggesting that there may be more as yet unidentified rare disorders that cause psychosis and that may be vitamin responsive. Identifying and patients with MTHFR deficiency polymorphisms along with 5q31.1 risk polymorphisms may provide a means of identifying some patients who may be responsive to specific combinations of vitamins and supplements.

It would be worth investigating further the nature of the relationship between phenylalanine hydroxylase, TP53 activity and/or the MTHFR C677T polymorphism. There does appear to be a relationship between the product of phenylalanine hydroxylase – tyrosine - and TP53 activity via tyrosine kinase and tyrosine hydroxylase. As well as a relationship between phenylalanine hydroxylase and folic acid.

A relationship between the proband's disorder and acetylcholine metabolism may account for the proband's responsiveness to lecithin which supports acetylcholine metabolism.

SNAP25 plays a key role in axonal growth and stimulates neurite sprouting. Axonal growth and neurite sprouting are thought to be linked to the causation of food intolerances. Also, the proband does carry SNAP25 SNPs associated with schizophrenia in IHDSF and he does report that his psychiatric symptoms are triggered by food intolerances and that his symptomology is significantly ameliorated by his elimination diet.

Whilst it does not follow that it must be the case that the proband's SNAP25 risk SNPs are the reason for this responsiveness, it could be the case that the proband's SNAP25 risk SNPs, whether alone or in combination with dysbindin risk SNPs, might result in axonal growth and neurite sprouting which could provide an explanation for some or all of the proband's food intolerances and for his dietary responsiveness.

In any event, SNAP25's connection to biogenic amines may provide an explanation for his amine intolerance. Perhaps SNAP25 modulates the disorder by impacting on amine signalling to the arginine/iNOS/NO pathway and the arginine/arginase/ornithine/polyamine pathway and negatively modulates the symptomology.

There are anecdotal reports of elimination diets resolving psychiatric symptoms including paranoia. However, there have been problems replicating such results. Identifying patients with known SNAP25 risk polymorphisms or SNAP25 risk SNPs combined with 5q31.1 risk SNPs and/or dysbindin risk SNPs may provide a means of identifying some patients who may be responsive.

In relation to the proband's calcium phosphate responsiveness, one or a combination of the factors listed above may provide an explanation for this responsiveness. Perhaps calcium phosphate modulates TP53 activity which in turn positively modulates the proband's symptomology.

It may be that SNPs carried by the proband at 5q31.1 impact on SAR1B function which results in his suffering from symptoms of Anderson's disease and that the group supplement's effectiveness in relation to the proband's *diarrhoea* was in part because the group supplement played a role in supporting his fatty acid metabolism. Alternatively or additionally, since potassium monophosphate may be connected to SAR1 and/or cyclic guanosine monophosphate activity and since early indications are that the proband's diarrhoea appears to be partially responsive to potassium chloride, it appears a possibility that the ingredient in the group supplement that the proband's diarrhoea responded so well to was potassium monophosphate.

It is unclear what role the proband's glutamic acid decarboxylase SNPs may play in relation to the proband's underlying disorder. It could be that they were inherited from the proband's mother and that they may provide an explanation for the proband's symptoms of anxiety and panic that are not common to the familial disorder on the paternal side. That they may be significant to the proband's supplement responsiveness is indicated by the suggestion that the proband's high urinary glutamic acid/creatinine ratio may correlate with when the proband was taking the group supplement. Performing a small scale trial to assess the efficacy of treatment of psychiatric patients who carry known GAD1 risk SNPs with the group supplement may be worthwhile.

If it is correct that the proband's disorder does result in elevated ornithine decarboxylase activity this could create conditions suitable to parasitic survival and thus an increased vulnerability to parasitic infections. If the proband does suffer from Blastocystisis Hominis infection and if Blastocystitis Hominis does down-regulate iNOS activity, this would run contrary to one of the ideas outlined in the hypothesis that there could be over-expression of a pathway from arginine to iNOS to cGMP in the proband. However, if iNOS is down-regulated in the proband this could again create conditions suitable to parasitic survival and thus vulnerability to parasitic infection; it could also cause excess arginine to be diverted away from iNOS to agmatine and it could still result in abnormal cGMP related activity as was previously proposed as part of the hypothesis.

If elevated agmatine levels create conditions suitable for the survival of fungi this could create a vulnerability to fungal infections.

If parasitic and/or fungal infections do occur in members of the population that carry SNPs that make them vulnerable to schizophrenia this might aggravate the underlying condition and/or act as a modifier or a trigger for episodes.

Along with their usual work up of psychiatric patients, which should include testing of B12 and other B vitamins, treating doctors should ensure that:

1. Patients are routinely tested for MTHFR risk polymorphisms;
2. Patients are routinely assessed to establish if they have any symptoms of food intolerance that warrant further investigations;

3. As part of their assessment of the patient they forensically question patients to obtain this information, given that failure to do this could result in the doctor missing food intolerance symptoms - for example, the proband had suffered severe diarrhoea from birth to about age 9 which then significantly improved and as a consequence the loose stools and less severe diarrhoea suffered by the proband thereafter were normalised in his mind;

4. They are fully up to date with the range of dermatological, gastric and nervous system symptoms that can be caused by food intolerances; and that

5. Patients are routinely tested for parasitic infections and that any identified parasitic infections are treated where appropriate.

It is interesting to note that some of the proband's abnormal results appear to be the opposite of those abnormal results that have been found in studies of schizophrenic patients.

DISCUSSION

Everyone is vulnerable to psychosis in the right set of circumstances, for example through extreme sleep deprivation. The proband's familial disorder, which appears to relate to schizophrenia in IHDSF and in which SNPs at 5q31.1 appear to play a central role, seems to make him significantly more vulnerable than most members of the population.

In the proband's case, it is hypothesised that other SNPs such as the MTHFR C677T (TT) polymorphism and SNAP25 risk polymorphisms negatively modulate the expression of his disorder resulting in his reaching criteria for psychosis. At best most of the treatments utilised by the proband target modulators of the disorder, such as the MTHFR deficiency; they do not resolve the underlying disorder.

There are probably a number of modulatory factors at play in most forms of schizophrenia. If sufficient modulatory factors can be identified and successfully treated in any schizophrenic patient this could be an effective way of treating the schizophrenia.

The idea that food intolerances could play such a significant role at least in some cases of schizophrenia is not easily accepted by the medical community. 'Where is the evidence?' is a cry which is often heard. However, there is plenty of evidence; the problem is that this evidence is anecdotal and not double blind within a controlled environment. The obvious difficulties of creating a controlled double blind test of dietary responsiveness are compounded by the fact that it may only be a small proportion of psychiatric patients who are responsive. However, there should be little doubt that some would be responsive. When considering this the reader should bear in mind that it has gradually become accepted that Celiac disease can result in mental health problems and slowly gluten intolerance in the absence of diagnosed Celiac disease is becoming accepted.

If starvation results in decreased jejunal villus heights and atrophy and if the proband's familial disorder results in increased jejunal villus heights, might this have provided a survival advantage in a population, such as the Irish, that was confronted with repeated famines and a very plain diet?

An alternative secondary hypothesis proposed that SNPs located at IL3/ACSL6 at 5q31.1 result in abnormal IL3 activity which in turn causes abnormal arginase activity which results in the same up-regulated and down-regulated pathways mentioned above.[75] Both this and the primary hypothesis in essence propose that a core defect closely connected to 5q31.1 and to IL3 and/or to ERE at this location are causative of the proband's disorder.

The following possibilities are consistent with both of these hypotheses:

1. The proband's family could carry a core defect that results in under-expression of the proline pathway which in turn impacts on activity along other connected pathways;

2. The carrying of this core defect could have resulted in evolutionary pressure causing sufferers to carry compensatory SNPs that up-regulate the proline pathway, particularly via TP53 function, and which compensate for disruption in the other connected pathways;

3. If the offspring of a sufferer does not carry the core defect, but does carry sufficient compensatory SNPs, this could result in massive over-expression of the proline

pathway which might in many cases be prenatally fatal at an early stage of gestation;

4. If the offspring of a sufferer does carry the core defect, but does not carry sufficient compensatory SNPs, this could result in overwhelming under-expression of the proline pathway which might in many cases be prenatally fatal at an early stage of gestation;

5. This might account for the high penetrance of the disorder in the proband's family;

6. If a surviving child of a sufferer does not carry the core defect but does carry at least some of the compensatory SNPs, these SNPs could still cause dysregulation of many of the same pathways outlined in this and the writer's previous paper;[76]

7. This type of dysregulation might in itself cause schizophrenia;

8. If there are sufficient familial disorders that directly impact on the relevant pathways and if as a result of evolutionary pressure these other familial disorders result in the carrying of compensatory SNPs that impact on the regulation of these pathways, the accumulation of a critical mass of these SNPs in a member of the general population could result in schizophrenia.

In which case, this might provide an explanation as to why genetic risk is conferred by a large number of alleles, including common alleles of small effect in the general population. In other words, the familial disorders that cause schizophrenia may play a significant role in producing the combinations of alleles that cause schizophrenia in the general population. However, it is noted that the third alternative hypothesis previously proposed is more obviously consistent with the prevailing view as to the cause of schizophrenia.[77]

Having said that, there is a means by which the first and second hypotheses may be unified. If IL3 plays a role in controlling ERE function then IL3 SNPs carried by the proband may be the proband's core defect and they may interrupt the presumed estradiol up-regulation of both P5CS and arginase.

Addendum 1

It has been reported that apolipoprotein B levels can be low in Anderson's disease[78] and that estrogen increases estrogen response elements transcription of the apolipoprotein gene in chickens.[79]

The proband's Apoliprotein B has now been tested and is mildly low (0.69g/l range 0.70-1.30).

The Apo B result is further evidence that the proband's familial disorder may be connected to Anderson's disease.

It was previously hypothesized that in the proband SNPs located at IL3 and/or ACSL6 at 5q31.1 result in or from the failure or partial failure of one or more ERE to activate a pathway leading from P5CS to proline. It might be that one or more of these ERE also play a role in activating SAR1B and or Apo B production.

Addendum 2

The proband has now tested positive for blastocystisis hominis.

Declaration

The writer has no formal medical training and is not a professional researcher.

Conflict of interest

Competing interest - none declared.

Ethics committee approval

This was not required because the proband is the writer.

References:

[1] Neville J. Analysing and attempting to connect the genetic and metabolic derangements underpinning a disorder which is linked to schizophrenia in Irish high density schizophrenia families. Medresind. 22.03.2015.

[2] Schizophrenia Working Group of the Psychiatric Genomics Consortium. Biological insights from 108 schizophrenia-associated genetic loci. Nature. 2014 Jul 24;511(7510):421-7. doi: 10.1038/nature13595. Epub 2014 Jul 22.

[3] Neville J. Analysing and attempting to connect the genetic and metabolic derangements underpinning a disorder which is linked to schizophrenia in Irish high density schizophrenia families. Medresind. 22.03.2015.

[4] Frosst P, Blom HJ, Milos R, Goyette P, Sheppard CA, Matthews RG, Boers GJ, den Heijer M, Kluijtmans LA, van den Heuvel LP, et al. A candidate genetic risk factor for vascular disease: a common mutation in methylenetetrahydrofolate reductase. Nat Genet. 1995 May;10(1):111-3.

[5] Zhou BS, Bu GY, Li M, Chang BG, Zhou YP. Tagging SNPs in the MTHFR gene and risk of ischemic stroke in a Chinese population. Int J Mol Sci. 2014 May 20;15(5):8931-40. doi: 10.3390/ijms15058931.

[6] Frosst P, Blom HJ, Milos R, Goyette P, Sheppard CA, Matthews RG, Boers GJ, den Heijer M, Kluijtmans LA, van den Heuvel LP, et al. A candidate genetic risk factor for vascular disease: a common mutation in methylenetetrahydrofolate reductase. Nat Genet. 1995 May;10(1):111-3.

[7]Frosst P, Blom HJ, Milos R, Goyette P, Sheppard CA, Matthews RG, Boers GJ, den Heijer M, Kluijtmans LA, van den Heuvel LP, et al. A candidate genetic risk factor for vascular disease: a common mutation in methylenetetrahydrofolate reductase. Nat Genet. 1995 May;10(1):111-3.

[8]Bönig H, Däublin G, Schwahn B, Wendel U. Psychotic symptoms in severe MTHFR deficiency and their successful treatment with betaine. Eur J Pediatr 2003 Mar;162(3):200-1. Epub 2003 Jan 18.

[9]Wang Q, Liu J, Liu YP, Li XY, Ma YY, Wu TF, Ding Y, Song JQ, Wang YJ, Yang YL. [Methylenetetrahydrofolate reductase deficiency-induced schizophrenia in a school-age boy]. Zhongguo Dan Dai Er Ke Za Shi. 2014 Jan;16(1):62-6. [Article in Chinese]

[10]Bönig H, Däublin G, Schwahn B, Wendel U. Psychotic symptoms in severe MTHFR deficiency and their successful treatment with betaine. Eur J Pediatr 2003 Mar;162(3):200-1. Epub 2003 Jan 18.

[11]Gilbody S, Lewis S, Lightfoot T. Methylenetetrahydrofolate reductase (MTHFR) genetic polymorphisms and psychiatric disorders: a HuGE review. AMJ Epidemiol. 2007 Jan 1;165(1):1-13. Epub 2006 Oct 30.

[12]Saetre P, Grove J, Børglum AD, Mors O, Werge T, Andreassen OA, Vares M, Agartz I, Terenius L, Jönsson EG. Methylenetetrahydrofolate reductase (MTHFR) C677T polymorphism and age at onset of schizophrenia: no consistent evidence for an association in the Nordic population. Am J Med Genet B neuropsychiatr Genet. 2012 Dec;159B(8):981-6. doi: 10.1002/ajmg.b.32104. Epub 2012 Oct 17.

[13]Bishop L, Kanoff R, Charnas L, Krenzel C, Berry SA, Schimmenti LA. Severe methylenetetrahydrofolate reductase (MTHFR)

deficiency: a case report of nonclassical homocystinuria. J Child Nerol. 2008 Jul;23(7):823-8. doi: 10.1177/0883073808315410.

[14]Diekman EF, de Koning TJ, Verhoeven-Duif NM, Rovers MM, van Hasselt PM. Survival and psychomotor development with early betaine treatment in probands with severe methylenetetrahydrofolate reductase deficiency. JAMA Nerol. 2014 Feb;71(2):188-94. doi: 10.1001/jamaneurol.2013.4915.

[15]Wang Q, Liu J, Liu YP, Li XY, Ma YY, Wu TF, Ding Y, Song JQ, Wang YJ, Yang YL. [Methylenetetrahydrofolate reductase deficiency-induced schizophrenia in a school-age boy]. Zhongguo Dan Dai Er Ke Za Shi. 2014 Jan;16(1):62-6. [Article in Chinese].

[16]Diekman EF, de Koning TJ, Verhoeven-Duif NM, Rovers MM, van Hasselt PM. Survival and psychomotor development with early betaine treatment in probands with severe methylenetetrahydrofolate reductase deficiency. JAMA Nerol. 2014 Feb;71(2):188-94. doi: 10.1001/jamaneurol.2013.4915.

[17]Y Ding, C-L Sun, L Li, M Li, L Francisco, M Sabado, B Hahn , J Gyorffy , J Noe, G P Larson, S J Forman, R Bhatia and S Bhatia. Genetic susceptibility to therapy-related leukemia after Hodgkin lymphoma or non-Hodgkin lymphoma: role of drug metabolism, apoptosis and DNA repair. Citation: Blood Cancer Journal (2012) 2, e58; doi:10.1038/bcj.2012.4. Published online 2 March 2012.

[18]Kuo CS, Huang CY, Kuo HT, Cheng CP, Chen CH, Lu CL, Yang FL, Syu Huang RF. Interrelationships among genetic C677T polymorphism of 5,10-methylenetetrahydrofolate reductase, biochemical folate status, and lymphocytic p53 oxidative damage in association with tumor malignancy and survivals of probands with hepatocellular carcinoma Mol Nutr Food Res. 2014 Feb;58(2):329-42. doi: 10.1002/mnfr.201200479. Epub

2013 Aug 29.

[19]Ulrich CM, Curtin K, Samowitz W, Bigler J, Potter JD, Caan B, Slattery ML. MTHFR variants reduce the risk of G:C->A:T transition mutations within the p53 tumor suppressor gene in colon tumors. J Nutr. 2005 Oct;135(10):2462-7.

[20]Kevere L, Purvina S, Bauze D, Zeibarts M, Andrezina R, Piekuse L, Brekis E, Purvins I. Homocysteine and MTHFR C677T polymorphism in children and adolescents with psychotic and mood disorders. Nord J Psychiatry. 2014 Feb; 68(2):129-36. doi: 10.3109/08039488.2013.782066. Epub 2013 Apr 16.

[21]Dudman NP, Tyrrell PA, Wilcken DE. Homocysteinemia: depressed plasma serine levels. Metabolism. 1987 Feb;36(2):198-201.

[22]Bishop L, Kanoff R, Charnas L, Krenzel C, Berry SA, Schimmenti LA. Severe methylenetetrahydrofolate reductase (MTHFR) deficiency: a case report of nonclassical homocystinuria. J Child Nerol. 2008 Jul;23(7):823-8. doi: 10.1177/0883073808315410.

[23]Diekman EF, de Koning TJ, Verhoeven-Duif NM, Rovers MM, van Hasselt PM Survival and psychomotor development with early betaine treatment in probands with severe methylenetetrahydrofolate reductase deficiency. JAMA Nerol. 2014 Feb;71(2):188-94. doi: 10.1001/jamaneurol.2013.4915.

[24]Wang Q, Liu J, Liu YP, Li XY, Ma YY, Wu TF, Ding Y, Song JQ, Wang YJ, Yang YL. Methylenetetrahydrofolate reductase deficiency-induced schizophrenia in a school-age boy. Zhongguo Dan Dai Er Ke Za Shi. 2014 Jan;16(1):62-6. [Article in Chinese].

[25]Kim. 2008. Folic Acid Supplementation and Cancer Risk: Point. doi: 10.1158/1055-9965.EPI-07-2557 Cancer Epidemiol

Biomarkers Prev September 2008 17; 2220.

[26]Talkowski ME, McClain L, Allen T, Bradford LD, Calkins M, Edwards N, Georgieva L, Go R, Gur R, Gur R, Kirov G, Chowdari K, Kwentus J, Lyons P, Mansour H, McEvoy J, O'Donovan MC, O'Jile J, Owen MJ, Santos A, Savage R, Toncheva D, Vockley G, Wood J, Devlin B, Nimgaonkar VL. Convergent patterns of association between phenylalanine hydroxylase variants and schizophrenia in four independent samples. Am J Med Genet B Neuropsychiatr Genet 2009 Jun 5;150B(4):560-9. doi: 10.1002/ajmg.b.30862.

[27]Bergen SE, Fanous AH, Walsh D, O'Neill FA, Kendler KS. Polymorphisms in SLC6A4, PAH, GABRB3, and MAOB and modification of psychotic disorder features. Schizophr Res. 2009 Apr;109(1-3):94-7. doi: 10.1016/j.schres.2009.02.009. Epub 2009 Mar 5.

[28]Goodfriend TL, Kaufman. Phenylalanine metabolism and folic acid antagonists. J Clin Invest. 1961 Sep;40:1743-50.

[29]Tsakiris S, Schulpis KH, Papaconstantinou ED, Tsakiris T, Tjamouranis I, Giannoulia-Karantana A Erythrocyte membrane acetylcholinesterase activity in subjects with MTHFR 677C-->T genotype. Clin Chem Lab Med. 2006;44(1):23-7.

[30]Pochini L, Scalise M, Galluccio M, Pani G, Siminovitch KA, Indiveri C. The human OCTN1 (SLC22A4) reconstituted in liposomes catalyzes acetylcholine transport which is defective in the mutant L503F associated to the Crohn's disease. Biochim Biophys Acta. 2012 Mar;1818(3):559-65. doi: 10.1016/j.bbamem.2011.12.014. Epub 2011 Dec 21.

[31]Osen-Sand A, Catsicas M, Staple JK, Jones KA, Ayala G, Knowles J, Grenningloh G, Catsicas S. Inhibition of axonal growth by SNAP-25 antisense oligonucleotides in vitro and in vivo. Nature.

1993 Jul 29;364(6436):445-8.

[32]Kimura K, Mizoguchi A. Regulation of growth cone extension by SNARE proteins   Ide CJ Histochem Cytochem. 2003 Apr;51(4):429-33.

[33]Fanous AH, Zhao Z, van den Oord EJ, Maher BS, Thiselton DL, Bergen SE, Wormley B, Bigdeli T, Amdur RL, O'Neill FA, Walsh D, Kendler KS, Riley BP.   Association study of SNAP25 and schizophrenia in Irish family and case-control samples. AMJ Med Genet B Neuropsychiatr Genet. 2010 Mar 5;153B(2):663-74. doi: 10.1002/ajmg.b.31037.

[34]Numakawa T, Yagasaki Y, Ishimoto T, Okada T, Suzuki T, Iwata N, Ozaki N, Taguchi T, Tatsumi M, Kamijima K, Straub RE, Weinberger DR, Kunugi H, Hashimoto R.   Evidence of novel neuronal functions of dysbindin, a susceptibility gene for schizophrenia. Hum Mol Genet. 2004 Nov 1;13(21):2699-708. Epub 2004 Sep 2.

[35]Kumamoto N, Matsuzaki S, Inoue K, Hattori T, Shimizu S, Hashimoto R, Yamatodani A, Katayama T, Tohyama M. Hyperactivation of midbrain dopaminergic system in schizophrenia could be attributed to the down-regulation of dysbindin. Biochem Biophys Res Commun.   2006 Jun 30;345(2):904-9. Epub 2006 May 6.

[36]Jia JM, Hu Z, Nordman J, Li Z. The schizophrenia susceptibility gene dysbindin regulates dendritic spine dynamics. J Neurosci. 2014 Oct 8;34(41):13725-36. doi: 10.1523/JNEUROSCI.0184-14.2014.

[37]Graff L, Castrop F, Bauer M, Höfler H, Gratzl M. Expression of vesicular monoamine transporters, synaptosomal-associated protein 25 and syntaxin1: a signature of human small cell lung

carcinoma. Cancer Res. 2001 Mar 1;61(5):2138-44.

[38]Neville J. Analysing and attempting to connect the genetic and metabolic derangements underpinning a disorder which is linked to schizophrenia in Irish high density schizophrenia families. Medresind. 22.03.2015.

[39]Madan E, Gogna R, Keppler B, Pati U. p53 increases intra-cellular calcium release by transcriptional regulation of calcium channel TRPC6 in GaQ3-treated cancer cells. PLoS One. 2013 Aug 16;8(8):e71016. doi: 10.1371/journal.pone.0071016. eCollection 2013.

[40]Lin J, Yang Q, Wilder PT, Carrier F, Weber DJ.. The calcium-binding protein S100B down-regulates p53 and apoptosis in malignant melanoma. J Biol Chem. 2010 Aug 27;285(35):27487-98. doi: 10.1074/jbc.M110.155382. Epub 2010 Jun 29.

[41] Alam MJ, Devi GR, Ravins, Ishrat R, Agarwal SM, Singh RK. Switching p53 states by calcium: dynamics and interaction of stress systems. Mol Biosyst. 2013 Mar;9(3):508-21. doi: 10.1039/c3mb25277a. Epub 2013 Jan 29.

[42]Corradini I, Verderio C, Sala M, Wilson MC, Matteoli M. SNAP-25 in neuropsychiatric disorders. Ann N Y Acad Sci. 2009 Jan;1152:93-9. doi: 10.1111/j.1749-6632.2008.03995.x.

[43]Zhang X, Kim-Miller MJ, Fukuda M, Kowalchyk JA, Martin TF. Ca2+-dependent synaptotagmin binding to SNAP-25 is essential for Ca2+-triggered exocytosis. Neuron. 2002 May 16;34(4):599-611.

[44]Gilks, William P.; Allott, Emma; Donohoe, Gary; Waddington, John L.; Gill, Michael; Corvin, Aiden P.; Morris, Derek W. The glutamatergic synapse protein HOMER2 is associated with

schizophrenia in the Irish population. Ulster Medical Journal;2008, Vol. 77 Issue 1, p66.

[45]Francis SH, Busch JL, Corbin JD, Sibley D. cGMP-dependent protein kinases and cGMP phosphodiesterases in nitric oxide and cGMP action. Pharmacol Rev. 2010 Sep;62(3):525-63. doi: 10.1124/pr.110.002907.

[46]Neville J. Analysing and attempting to connect the genetic and metabolic derangements underpinning a disorder which is linked to schizophrenia in Irish high density schizophrenia families (free access on preview). Medresind. 22.03.2015.

[47]Neville J. Analysing and attempting to connect the genetic and metabolic derangements underpinning a disorder which is linked to schizophrenia in Irish high density schizophrenia families (free access on preview). Medresind. 22.03.2015.

[48]Mattson MP Parkinson's disease: don't mess with calcium. J Clin Invest. 2012 Apr;122(4):1195-8. doi: 10.1172/JCI62835. Epub 2012 Mar 26.

[49]Hayssam Dannoura, Nathalie Berriot-Varoqueaux, Patricia Amati, Veronique badie, Nicole Verthier, Jacques Schmitz, John R. Wetterau, Marie-Elisabeth Samson-Bouma, Lawrence P. Aggerbeck Atherosclerosis and Lipoproteins. Anderson's Disease Exclusion of Apolipoprotein and Intracellular Lipid Transport Genes. Arteriosclerosis, Thrombosis, and Vascular Biology. 1999; 19: 2494-2508 doi: 10.1161/01.ATV.19.10.2494

[50]Neville J. Analysing and attempting to connect the genetic and metabolic derangements underpinning a disorder which is linked to schizophrenia in Irish high density schizophrenia families.

Medresind. 22.03.2015.

[51]Weissman JT, Plutner H, Balch WE. The mammalian guanine nucleotide exchange factor mSec12 is essential for activation of the Sar1 GTPase directing endoplasmic reticulum export. Traffic. 2001 Jul;2(7):465-75.

[52]McMahon C, Studer SM, Clendinen C, Dann GP, Jeffrey PD, Hughson FM. The structure of Sec12 implicates potassium ion coordination in Sar1 activation. J Biol Chem. 2012 Dec 21;287(52):43599-606. doi: 10.1074/jbc.M112.420141. Epub 2012 Oct 29.

[53]Hayssam Dannoura, Nathalie Berriot-Varoqueaux, Patricia Amati, Veronique badie, Nicole Verthier, Jacques Schmitz, John R. Wetterau, Marie-Elisabeth Samson-Bouma, Lawrence P. Aggerbeck Atherosclerosis and Lipoproteins. Anderson's Disease Exclusion of Apolipoprotein and Intracellular Lipid Transport Genes. Arteriosclerosis, Thrombosis, and Vascular Biology. 1999; 19: 2494-2508 doi: 10.1161/01.ATV.19.10.2494

[54]Hettema JM, An SS, Neale MC, Bukszar J, van den Oord EJ, Kendler KS, Chen X. Association between glutamic acid decarboxylase genes and anxiety disorders, major depression, and neuroticism. 2006 Aug;11(8):752-62. Epub 2006 May 23.

[55]Reynertson KA, Garay M, Nebus J, Chon S, Kaur S, Mahmood K, Kizoulis M, Southall MD. Anti-inflammatory activities of colloidal oatmeal (Avena sativa) contribute to the effectiveness of oats in treatment of itch associated with dry, irritated skin. J Drugs Dermatol. 2015 Jan;14(1):43-8.

[56]Neville J. Analysing and attempting to connect the genetic and metabolic derangements underpinning a disorder which is linked

to schizophrenia in Irish high density schizophrenia families. Medresind. 22.03.2015.

[57]Wu G, Flynn NE, Knabe DA. 2000.   Enhanced intestinal synthesis of polyamines from proline in cortisol-treated piglets. Am J Physiol Endocrinol Metab 279: E395–E402, 2000.

[58]Wirén M, Söderholm JD, Lindgren J, Olaison G, Permert J, Yang H, Larsson J. Effects of starvation and bowel resection on paracellular permeability in rat small-bowel mucosa in vitro. Scand J Gastroenterol. 1999 Feb;34(2):156-62.

[59]Neville J. Analysing and attempting to connect the genetic and metabolic derangements underpinning a disorder which is linked to schizophrenia in Irish high density schizophrenia families. Medresind. 22.03.2015.

[60]Bahman Khalili ، * ; Reza Imani ; and Sanaz Boostani ، Intestinal Parasitic Infections in Chronic Psychiatric Patients in Sina Hospital Shahre-Kord, Iran Jundishapur Journal of Microbiology. 2013 May; 6(3): 252-5. , DOI: 10.5812/jjm.5092

[61]Yolken RH[1], Dickerson FB, Fuller Torrey E. Toxoplasma and schizophrenia. Parasite Immunol. 2009 Nov;31(11):706-15. doi: 10.1111/j.1365-3024.2009.01131.x.

[62]Mirza H, Wu Z, Kidwai F, Tan KS. A metronidazole-resistant isolate of Blastocystis spp. is susceptible to nitric oxide and downregulates intestinal epithelial inducible nitric oxide synthase by a novel parasite survival mechanism. Infect Immun. 2011 Dec;79(12):5019-26. doi: 10.1128/IAI.05632-11. Epub 2011 Sep 19.

[63]Lim MX, Png CW, Tay CY, Teo JD, Jiao H, Lehming N, Tan KS, Zhang Y. Differential regulation of proinflammatory cytokine

expression by mitogen-activated protein kinases in macrophages in response to intestinal parasite infection. Infect Immun. 2014 Nov;82(11):4789-801. doi: 10.1128/IAI.02279-14. Epub 2014 Aug 25.

[64]Uzbay T, Goktalay G, Kayir H, Eker SS, Sarandol A, Oral S, Buyukuysal L, Ulusoy G, Kirli S. Increased plasma agmatine levels in patients with schizophrenia. J Psychiatr Res. 2013 Aug;47(8):1054-60. doi: 10.1016/j.jpsychires.2013.04.004. Epub 2013 May 7.

[65]Neville J. Analysing and attempting to connect the genetic and metabolic derangements underpinning a disorder which is linked to schizophrenia in Irish high density schizophrenia families. Medresind. 22.03.2015.

[66]Gardiner DM, Kazan K, Manners JM. Novel genes of Fusarium graminearum that negatively regulate deoxynivalenol production and virulence. Mol Plant Microbe Interact. 2009 Dec;22(12):1588-600. doi: 10.1094/MPMI-22-12-1588.

[67]Schneider E, Ihle JN, Dy M. Homogeneous interleukin 3 enhances arginase activity in murine hematopoietic cells. Lymphokine Res. 1985 Spring;4(2):95-102.

[68]Hayashi T, Esaki T, Sumi D, Mukherjee T, Iguchi A, Chaudhuri G. Modulating role of estradiol on arginase II expression in hyperlipidemic rabbits as an atheroprotective mechanism. Proc Natl Acad Sci U S A. 2006 Jul 5;103(27):10485-90. Epub 2006 Jun 26.

[69]Traish AM, Kim NN, Huang YH, Min K, Munarriz R, Goldstein I. Sex steroid hormones differentially regulate nitric oxide synthase and arginase activities in the proximal and distal rabbit vagina.. Int J Impot Res. 2003 Dec;15(6):397-404.

[70]Xiong Y, Yu Y, Montani JP, Yang Z, Ming XF. Arginase-II induces vascular smooth muscle cell senescence and apoptosis through p66Shc and p53 independently of its l-arginine ureahydrolase activity: implications for atherosclerotic plaque vulnerability. J Am Heart Assoc. 2013 Jul 5;2(4):e000096. doi: 10.1161/JAHA.113.000096.

[71]Rogers A, Eastell R. The effect of 17beta-estradiol on production of cytokines in cultures of peripheral blood. Bone. 2001 Jul;29(1):30-4.

[72]Knöferl MW, Jarrar D, Angele MK, Ayala A, Schwacha MG, Bland KI, Chaudry IH. 17 beta-Estradiol normalizes immune responses in ovariectomized females after trauma-hemorrhage. Am J Physiol Cell Physiol. 2001 Oct;281(4):C1131-8.

[73]Luo XJ, Li M, Huang L, Nho K, Deng M, Chen Q, Weinberger DR, Vasquez AA, Rijpkema M, Mattay VS, Saykin AJ, Shen L, Fernández G, Franke B, Chen JC, Chen XN, Wang JK, Xiao X, Qi XB, Xiang K, Peng YM, Cao XY, Li Y, Shi XD; Alzheimer's Disease Neuroimaging Initiative, Gan L, Su B. The interleukin 3 gene (IL3) contributes to human brain volume variation by regulating proliferation and survival of neural progenitors. PLoS One. 2012;7(11):e50375. doi: 10.1371/journal.pone.0050375. Epub 2012 Nov 30.

[74]Klinge CM. Estrogen receptor interaction with estrogen response elements.: Oxford Journals. Science & Mathematics. Nucleic Acids Research Volume 29, Issue 14. Pp. 2905-2919.

[75]Neville J. Analysing and attempting to connect the genetic and metabolic derangements underpinning a disorder which is linked to schizophrenia in Irish high density schizophrenia families. Medresind. 22.03.2015.

[76]Neville J. Analysing and attempting to connect the genetic and metabolic derangements underpinning a disorder which is linked to schizophrenia in Irish high density schizophrenia families. Medresind. 22.03.2015.

[77]Neville J. Analysing and attempting to connect the genetic and metabolic derangements underpinning a disorder which is linked to schizophrenia in Irish high density schizophrenia families. Medresind. 22.03.2015.

[78]Strich D, Goldstein R, Phillips A, Shemer R, Goldberg Y, Razin A, Freier S. Anderson's disease: no linkage to the apo B locus. J Pediatr Gastroenterol Nutr. 1993 Apr;16(3):257-64.

[79]Edinger RS, Mambo E, Evans MI. Estrogen-dependent transcriptional activation and vitellogenin gene memory. Mol Endocrinol. 1997 Dec;11(13):1985-93.

THE POCKET GUIDE FOR

NERVOUS

NET
WORK
ERS

Ash Mashhadi

Illustrations by Casper Mashhadi

Copyright © 2016 Ash Mashhadi
Illustrations copyright © 2016 Casper Mashhadi

The author has asserted his rights under the Copyright,
Designs and Patents Act, 1988 to be identified as the
author of this work.

First published in Great Britain by Design Inspiration

Designed and typeset by Design Inspiration
Inspiration.co.uk

A CIP catalogue record for this book is available from the
British Library.

ISBN 978-1-326-59439-8